my itty-bitty bio

Greta Thunberg

CHERRY LAKE PRESS

Published in the United States of America by Cherry Lake Publishing Group
Ann Arbor, Michigan
www.cherrylakepublishing.com

Reading Adviser: Beth Walker Gambro, MS, Ed., Reading Consultant, Yorkville, IL
Book Designer: Jennifer Wahi
Illustrator: Jeff Bane

Photo Credits: © JJFarq/shutterstock, 5; ©Photo by Bengt Nyman/Wikimedia, 7; ©NASA Images/Shutterstock, 9; © Antonello Marangi/Dreamstime, 11; ©Alexandros Michailidis/Shutterstock, 13; © Per Grunditz/Dreamstime, 15; © Liv Oeian/Shutterstock, 17, 22; © Franz1212/Dreamstime, 19, 23; ©Daniele COSSU/Shutterstock, 21; Cover, 1, 10, 12, 16, Jeff Bane; Various frames throughout, Shutterstock

Copyright ©2022 by Cherry Lake Publishing Group
All rights reserved. No part of this book may be reproduced or utilized in
any form or by any means without written permission from the publisher.

Cherry Lake Press is an imprint of Cherry Lake Publishing Group.

Library of Congress Cataloging-in-Publication Data

Names: Sarantou, Katlin, author. | Bane, Jeff, 1957- illustrator.
Title: Greta Thunberg / Katlin Sarantou ; illustrated by Jeff Bane.
Description: Ann Arbor, Michigan : Cherry Lake Publishing, 2021. | Series: My itty-bitty bio | Includes index.
Identifiers: LCCN 2021007982 (print) | LCCN 2021007983 (ebook) | ISBN 9781534186880 (hardcover) | ISBN 9781534188280 (paperback) | ISBN 9781534189683 (pdf) | ISBN 9781534191082 (ebook)
Subjects: LCSH: Thunberg, Greta, 2003---Juvenile literature. | Child environmentalists--Sweden--Biography--Juvenile literature. | Environmentalists--Sweden--Biography--Juvenile literature.
Classification: LCC GE56.T58 S26 2021 (print) | LCC GE56.T58 (ebook) | DDC 363.70092 [B]--dc23
LC record available at https://lccn.loc.gov/2021007982
LC ebook record available at https://lccn.loc.gov/2021007983

Printed in the United States of America
Corporate Graphics

table of contents

My Story 4

Timeline 22

Glossary 24

Index 24

About the author: Katlin Sarantou grew up in the cornfields of Ohio. She enjoys reading and dreaming of faraway places.

About the illustrator: Jeff Bane and his two business partners own a studio along the American River in Folsom, California, home of the 1849 Gold Rush. When Jeff's not sketching or illustrating for clients, he's either swimming or kayaking in the river to relax.

my story

My name is Greta Thunberg.
I was born January 3, 2003,
in Sweden.

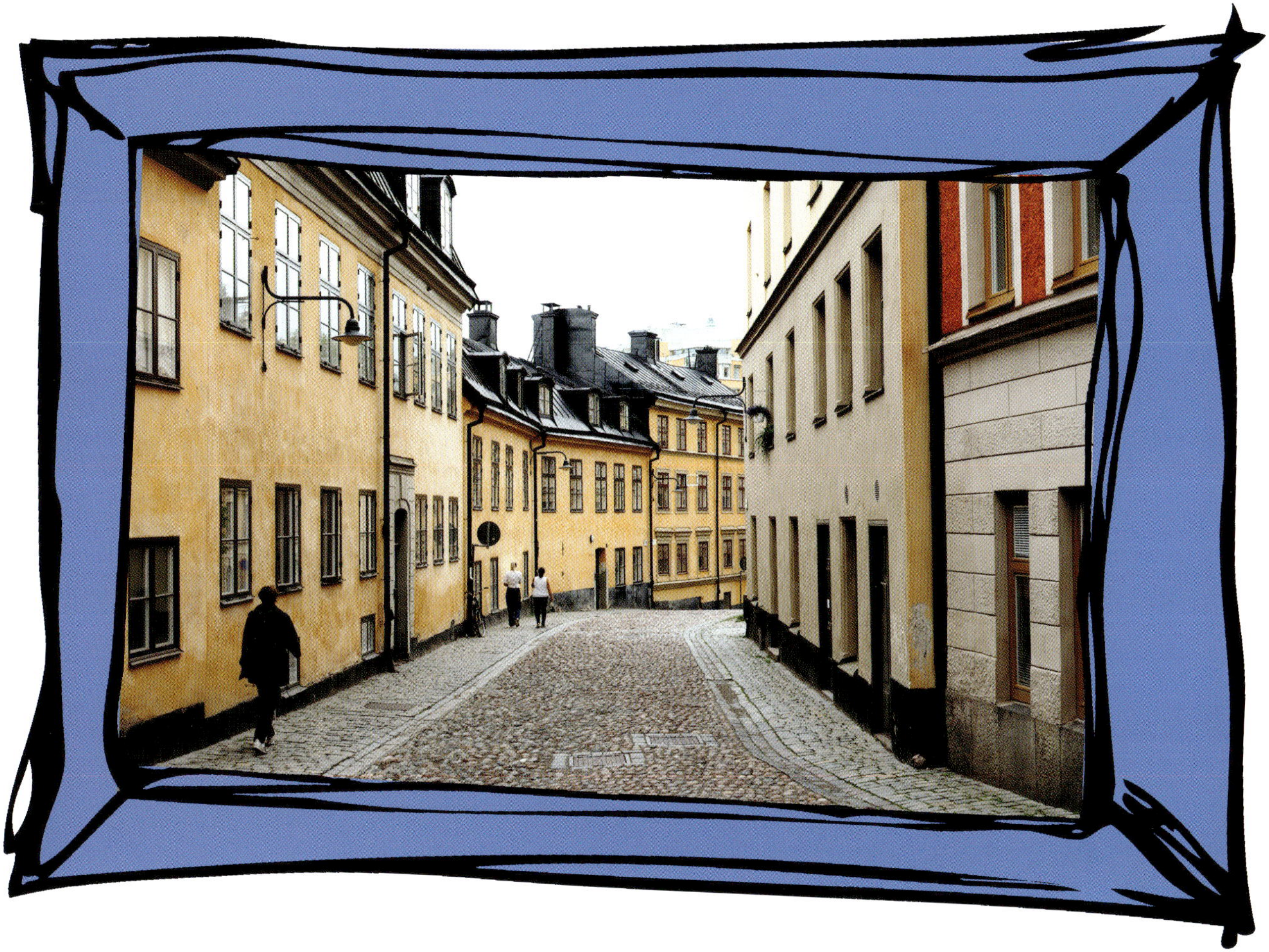

My father is an actor. My mother is an opera singer.

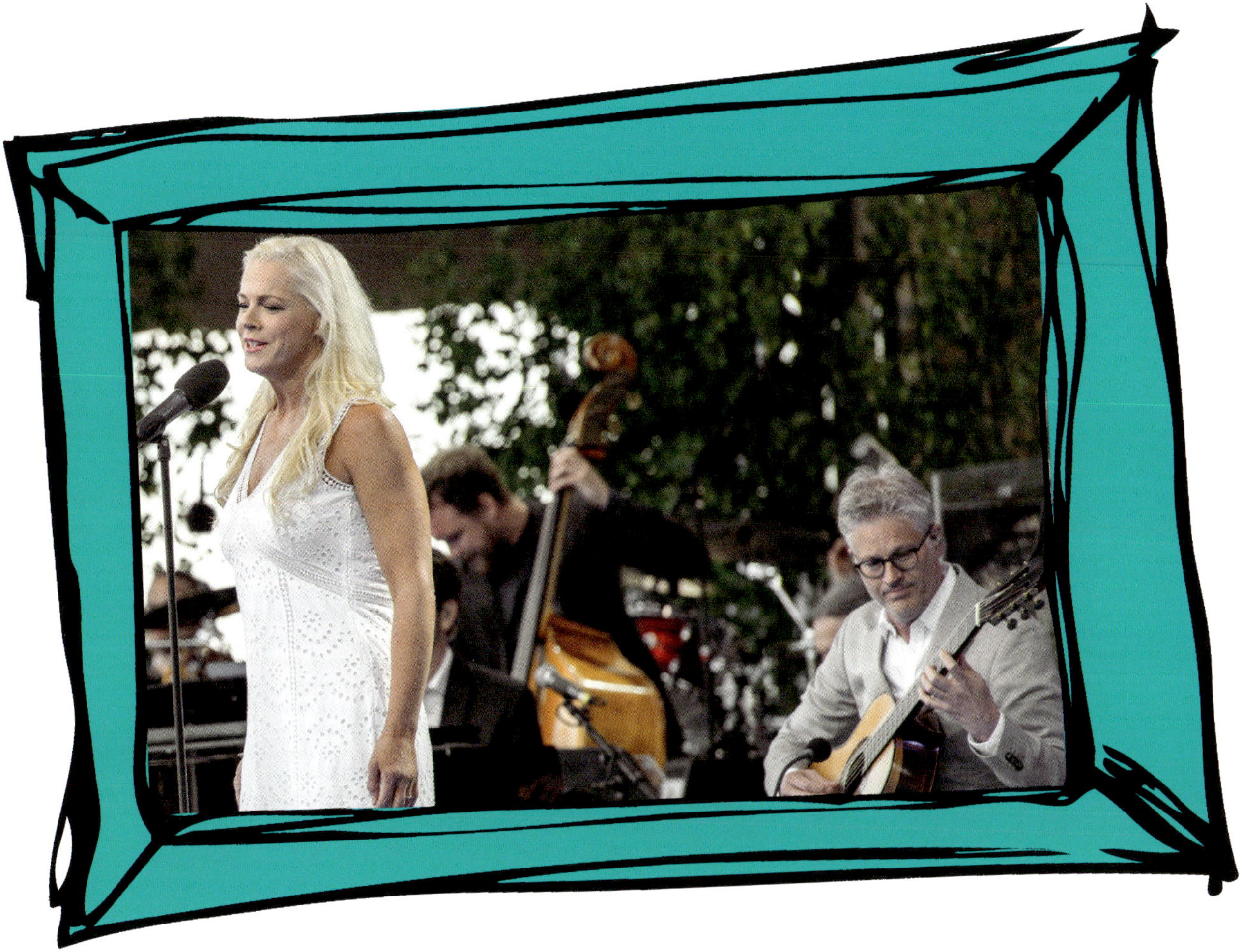

I first heard about **climate change** when I was 8 years old. I wanted to help.

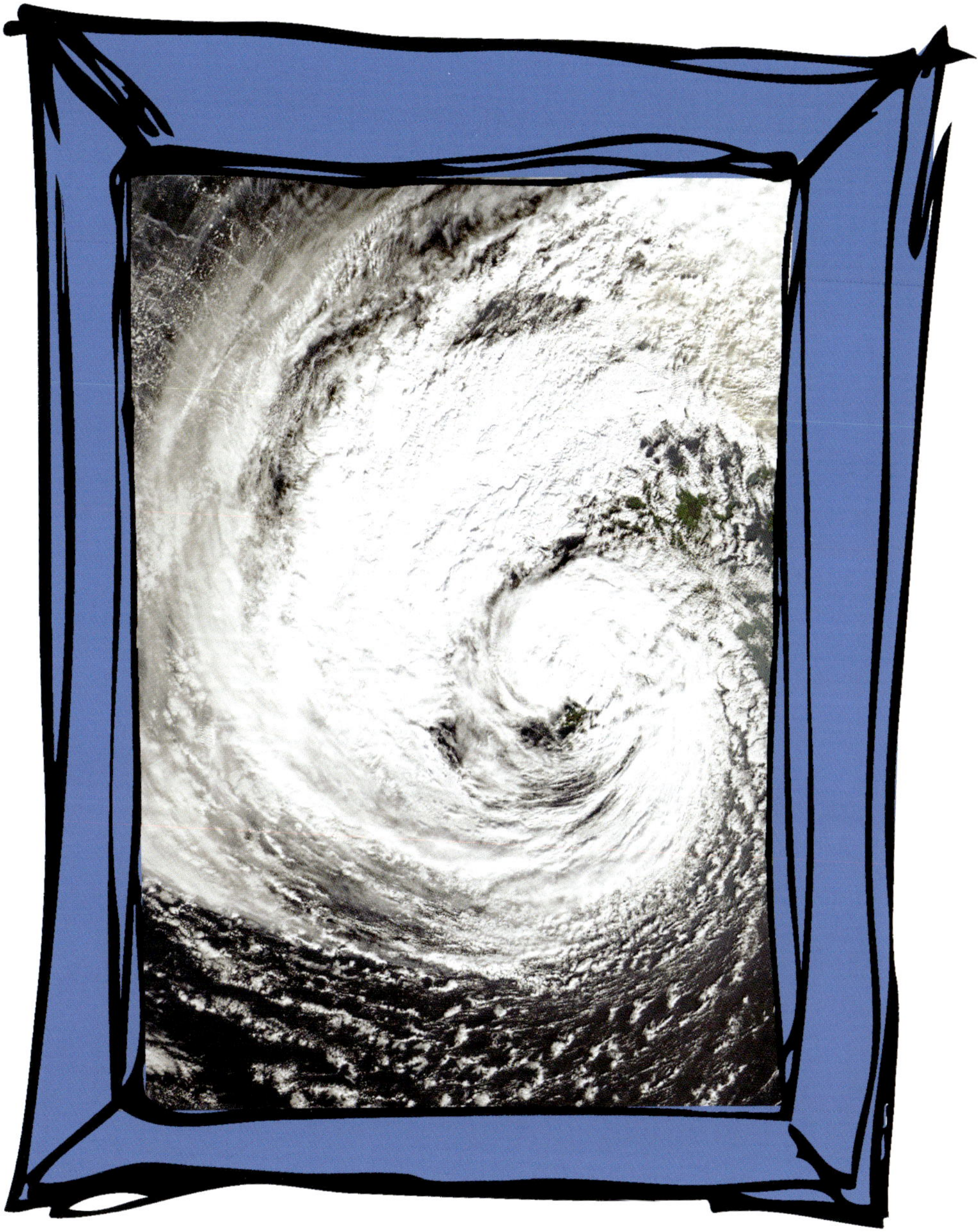

I have **ASD**.

I don't let it stop me. I think of it like a superpower.

What is your superpower?

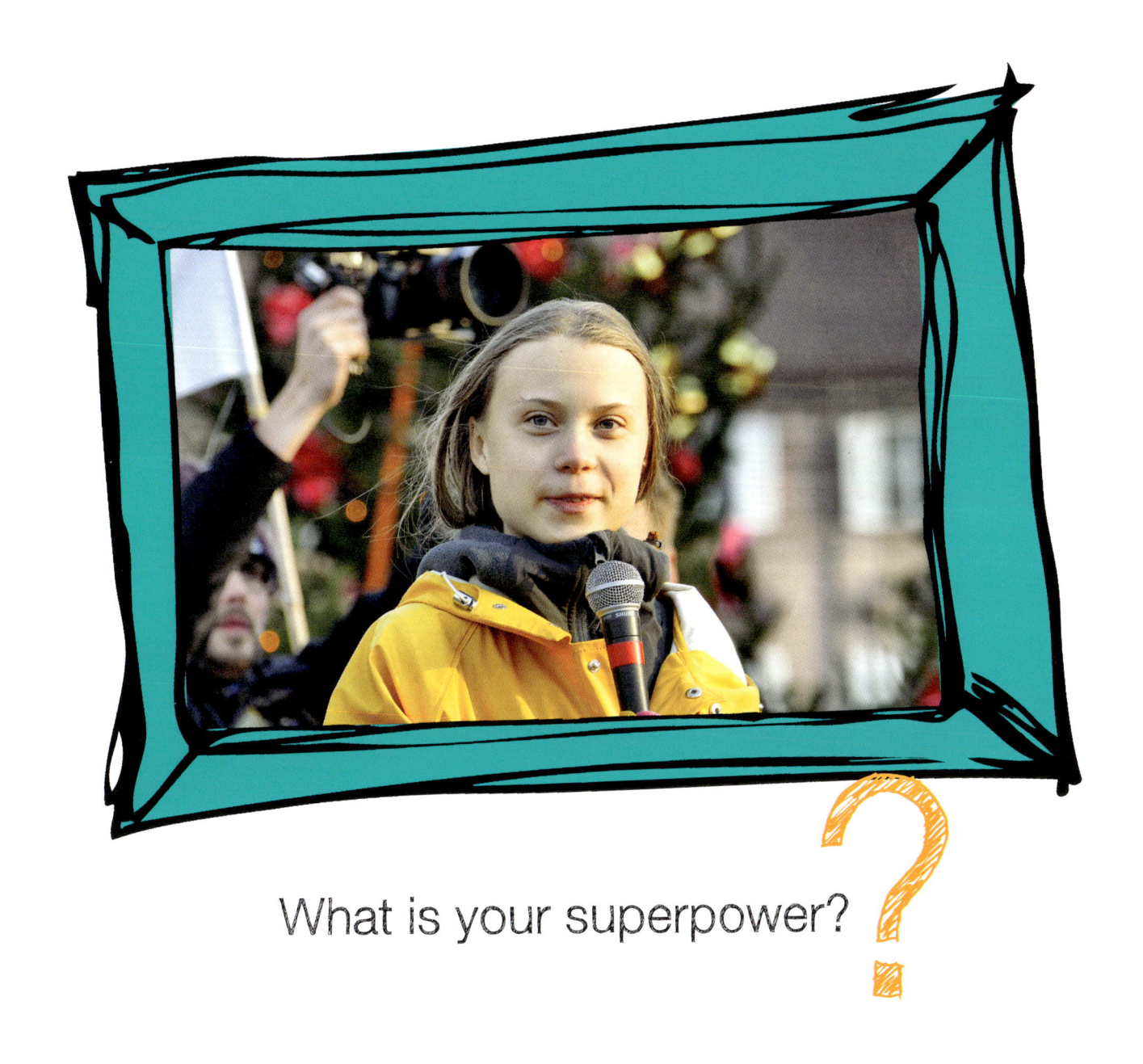

Sometimes I have **anxiety**.
I only speak when it's important.
I speak about the environment.

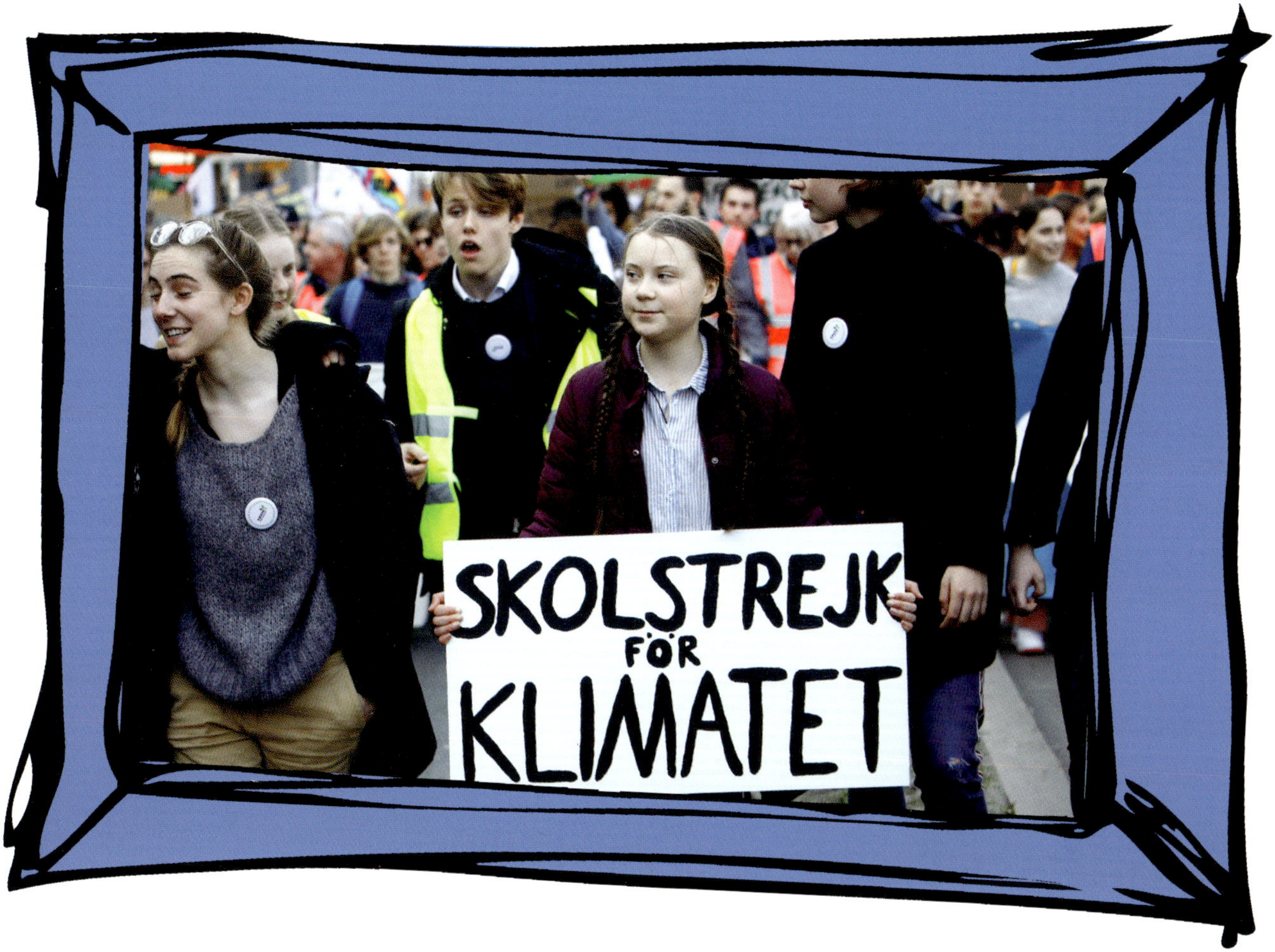
SKOLSTREJK
FÖR
KLIMATET

I helped my parents lower our **carbon footprint**.

What can you do at home to help the environment?

I began public speaking in 2018.
I started school climate **strikes**.
I inspired other kids.

COOL KIDS
SAVING A
HOT PLANET
JORDEN KAN
LEVA LÄNGRE
OM DET HÄNDER NÅGOT NU!
TIME IS
RUNNING
SKOLSTREJK
FÖR
NOT THE
HAR
FEBER

I was voted *Time* magazine Person of the Year in 2019. I am the youngest person to get this award. I was **nominated** twice for the Nobel Peace Prize.

What do you like to do?

I am an environmental **activist**. I challenge world leaders to do better.

What would you like to ask me?

timeline

2018

2000

Born 2003

22

2019

2100

glossary & index

glossary

activist (AK-tih-vist) a person supporting a cause

anxiety (ang-ZYE-uh-tee) nervousness or fear

ASD (AY ESS DEE) stands for autism spectrum disorder, which can involve limited social skills and repeated behaviors

carbon footprint (KAR-buhn FUT-print) the amount of gases released into the air due to a person's lifestyle

climate change (KLYE-mit CHAYNJ) a significant change in the planet's climate and weather

nominated (NAH-muh-nayt-uhd) to be proposed for an honor

strikes (STRYKES) temporary stops of something in protest

index

activist, 20
anxiety, 12
ASD, 10
award, 18
carbon footprint, 14
climate, 8, 16

environment, 12, 15, 20

Nobel Peace Prize, 18

superpower, 10, 11
Sweden, 4

Time magazine, 18